ALTERNATOR BOOKS™

ENERGY

INVESTIGATIONS

KAREN LATCHANA KENNEY

To our future scientists and their unknown discoveries

Content consultant: Kevin Finerghty, adjunct professor of Geology at State University of New York, Oswego; Earth science teacher at Pulaski Academy and Central Schools, Pulaski, New York

Lerner Publications Company
A division of Lerner Publishing Group, Inc.
241 First Avenue North
Minneapolis, MN 55401 USA

For reading levels and more information, look up this title at www.lernerbooks.com.

Main body text set in Aptifer Slab Regular 11.5/18.
Typeface provided by Linotype AG.

Library of Congress Cataloging-in-Publication Data

Names: Kenney, Karen Latchana.
Title: Energy investigations / Karen Latchana Kenney.
Description: Minneapolis : Lerner Publications, [2018] | Series: Key questions in physical science | Audience: Age 8–12. | Audience: Grade 4 to 6. | Includes bibliographical references and index.
Identifiers: LCCN 2016043840 (print) | LCCN 2016045066 (ebook) | ISBN 9781512440034 (lb : alk. paper) | ISBN 9781512449563 (eb pdf)
Subjects: LCSH: Power resources—Juvenile literature. | Power resources—Experiments—Juvenile literature. | Solar energy—Juvenile literature. | Energy metabolism—Juvenile literature.
Classification: LCC TJ163.23 .K4525 2018 (print) | LCC TJ163.23 (ebook) | DDC 530—dc23

LC record available at https://lccn.loc.gov/2016043840

Manufactured in the United States of America
1-42265-26122-3/2/2017

CONTENTS

DROPPING IN

You're standing on a skateboard, about to drop into a half-pipe. You lean forward, your skateboard's nose drops, and you fly down the wall of the ramp. You reach the bottom and begin gliding up the other wall. Soon you're coasting back and forth, riding the walls.

Have you ever wondered how a skateboarder speeds up and down a half-pipe? Or why sunlight feels warm? Or why your body needs food? It all has to do with energy,

Energy is what allows a skateboarder to continue moving up and down a half-pipe.

Sunlight is one of the most important kinds of energy. Sunlight gives plants energy and supports life on Earth.

or the ability to do work. Over the centuries, scientists have wondered about energy. They've formed theories and tested their ideas to come up with answers based on evidence. This process is known as scientific inquiry. Sometimes it took scientists years of research, lots of experiments, and even a few wrong ideas to answer their questions about energy. First, they noticed the energy of motion. Then they discovered that there's energy in heat and sunlight. Later, scientists learned about the energy of electricity and fuel. Through questions, tests, and theories, scientists have discovered a lot about how energy powers our world.

CHAPTER 1

WHERE DOES ENERGY GO?

Skateboarding on a half-pipe uses two kinds of energy. At the top of the ramp, a skateboarder has lots of potential energy, meaning she is in a position where she can gain energy. Once the skateboarder drops, kinetic energy, or the energy of motion, is being used. A skateboarder rocks between potential and kinetic energy while riding the walls of a half-pipe.

Hermann von Helmholtz was one of the first scientists to write about the conservation of energy. He came up with this idea while trying to understand how animals create heat.

EARLY ENERGY DISCOVERIES

But how exactly does that work? How is energy used, and where does it go? In the nineteenth century, three scientists were trying to figure out where potential energy goes. Through their studies, German scientists Hermann von Helmholtz and Julius Robert von Mayer and British scientist James Prescott Joule all concluded that energy is **conserved**—it transfers from one form to another but is never destroyed.

That's what happens when a skateboarder flies up and down a half-pipe. The energy is never used up. It simply changes from one form to another. This idea became known as the conservation of energy. However, forces such as friction will eventually work against the skateboarder, slowing the skateboard down and taking its energy away.

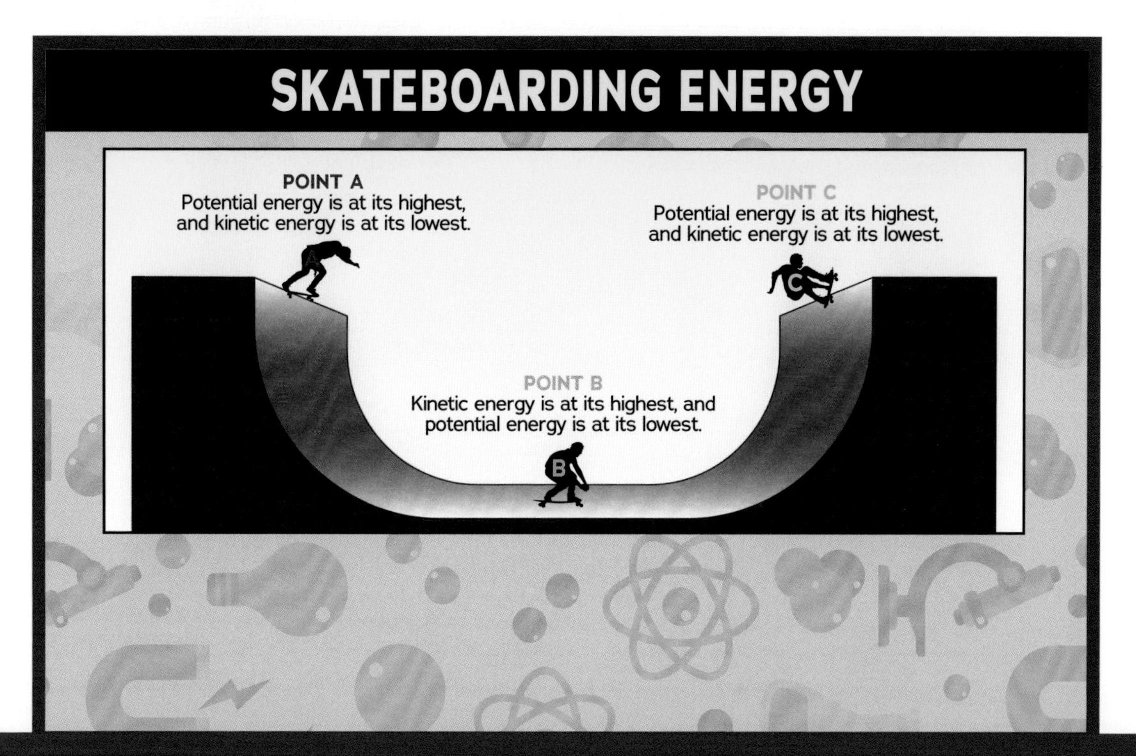

SCIENCE IN PRACTICE

In the early eighteenth century, French scientist Émilie du Châtelet tried an experiment to test the energy of a falling object. She dropped lead balls onto a soft clay floor. When the balls landed, they dented the clay. She dropped the balls from different heights and tested how the speed of the balls affected the size of the dents. Balls moving faster made larger dents. A ball moving twice as fast as another made a dent four times as large. A ball moving three times as fast made a dent nine times as large. These results contributed to scientists' understanding of kinetic energy and the conservation of energy.

Du Châtelet's experiment proved one of Isaac Newton's theories wrong. He thought a ball moving twice as fast would make a dent twice as large.

Potential energy and kinetic energy are the two main kinds of energy. And each of these comes in many more forms. Potential energy can be locked inside a plant's atoms as chemical energy. It's inside an atom's nucleus as nuclear energy. And it's the tension in a stretched rubber band about to snap. Kinetic energy is in the booming sound from fireworks or the light from the sun. It's the motion of objects, like a fast car. And it's also electricity, like the flash of a lightning bolt.

A lightning strike has enough energy to power a house for a month.

CHAPTER 2

WHAT ENERGY IS IN SUNLIGHT?

One form of kinetic energy can be felt on a sunny day. You might get warm and start to sweat as you feel the sun's heat. As ultraviolet rays kick into action, you might even begin to get a sunburn. What you're feeling and seeing all comes from the sun's radiant energy.

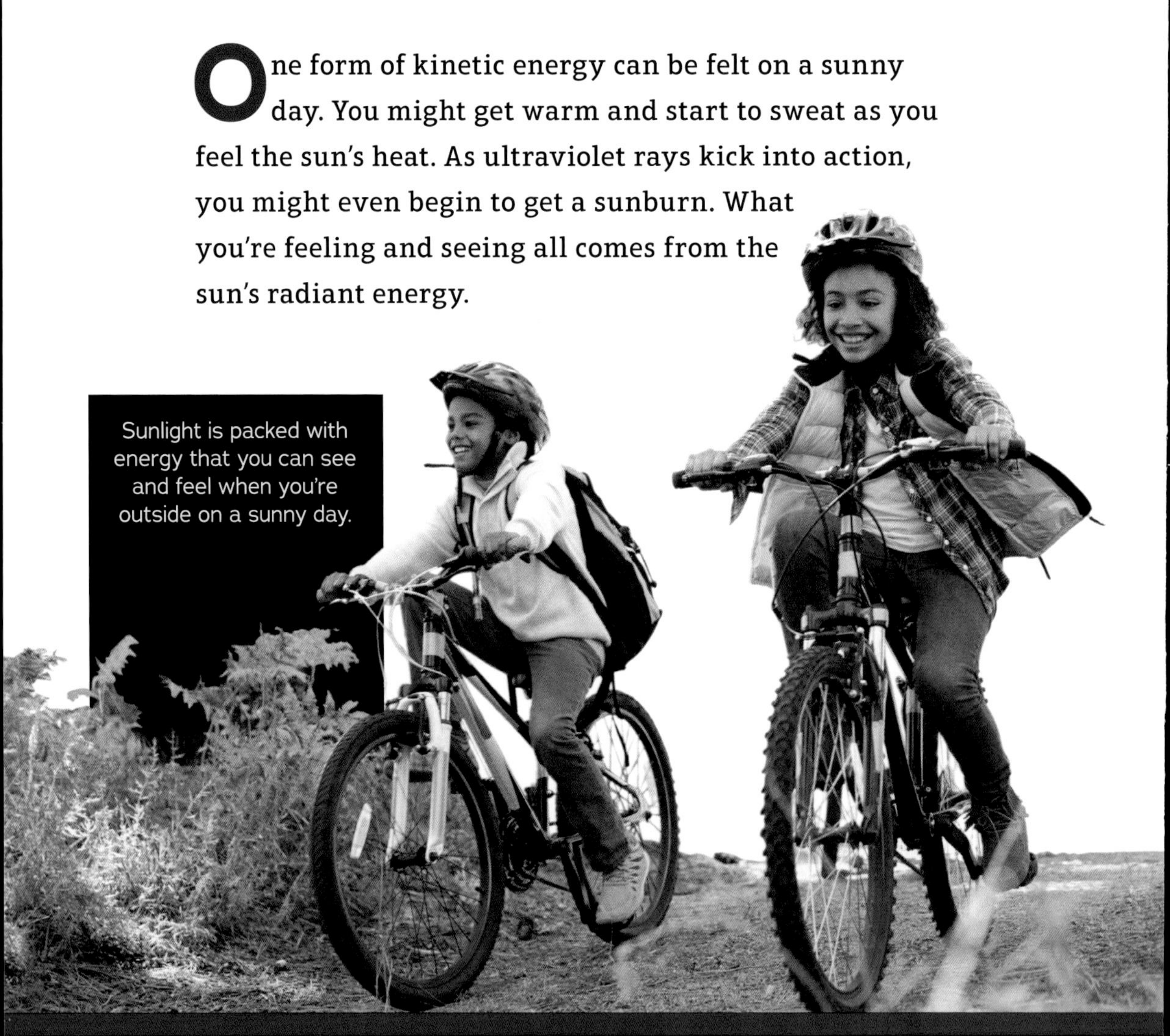

Sunlight is packed with energy that you can see and feel when you're outside on a sunny day.

You can see light from the sun and feel its heat, but there's much more to sunlight. It is made up of energy known as **electromagnetic radiation**. This energy includes not only ultraviolet and visible light but also light waves called gamma rays, X-rays, infrared, microwave, and radio waves.

EARLY ENERGY DISCOVERIES

The sun is the original source of almost all the energy on Earth. In the nineteenth century, people began to wonder if the sun's energy could be used to power machines.

Our eyes can only see some of the light waves given off by the sun. This visible light is made up of all the colors of the rainbow.

SCIENCE IN PRACTICE

French inventor Augustin Mouchot wanted to create a solar-powered engine in the nineteenth century. In 1861 he experimented by pouring water into an iron bucket surrounded by sun **reflectors**. When the water heated, it made steam that powered a small motor. In 1878 he showed his solar-powered machines at a fair in Paris that displayed the latest technology from around the world. Mouchot even used solar power to run an ice-making machine.

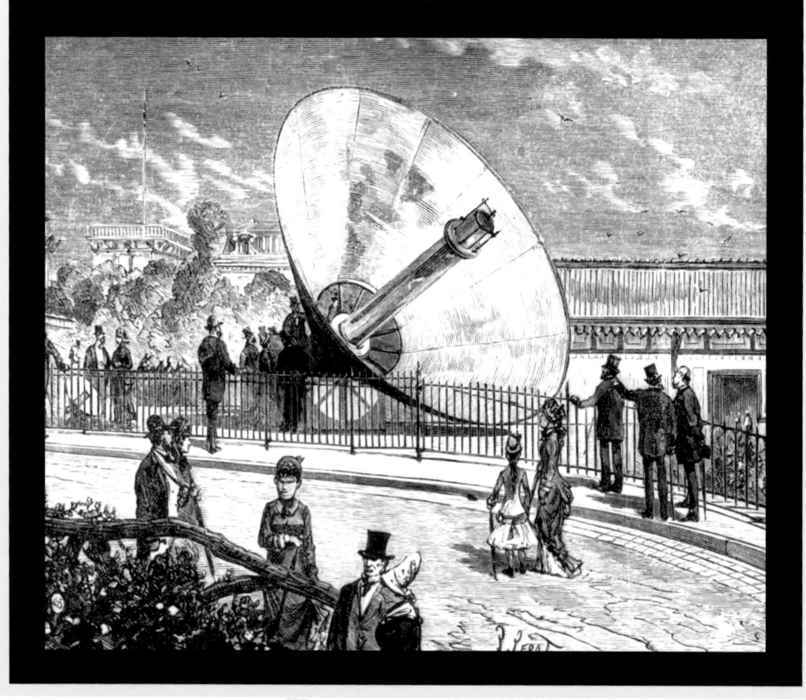

Mouchot worked for six years to develop a way to collect the sun's energy.

In 1839 nineteen-year-old French physicist Alexandre Edmond Becquerel was the first to find out. He experimented with platinum, a metal that conducts electricity, and different types of light. Becquerel noticed that the metal generated an electric current when it was exposed to light. It was easiest to generate electricity using sunlight or blue light. He also experimented with different metal and discovered that silver could also generate electricity. Becquerel had just discovered the photovoltaic effect, the changing of particles of light, called photons, into electricity.

Becquerel's discovery eventually led to the solar cell, a device that uses the photovoltaic effect to convert the sun's energy into electricity. These days, solar cells power many things, from calculators to homes to the International Space Station.

The first successful solar panel and solar battery were installed in 1955 for a telephone system in Georgia.

CHAPTER 3

HOW DOES YOUR BODY GET ENERGY?

Much of the sun's energy is blocked by Earth's atmosphere, but visible light does reach Earth's surface. The energy in this light is what makes life on our planet possible. Just take a bite out of a cucumber or an apple, and you'll get some of the sun's energy. Plants store the sun's energy in a form your body can use.

FINDING PHOTOSYNTHESIS

In the late eighteenth century, doctor and scientist Jan Ingenhousz experimented with plants and light to understand how plants collect and store the sun's energy. To conduct

Without enough sunlight, plants do not receive the energy they need to continue growing.

his experiment, Ingenhousz placed plants underwater in the sunlight and then again in the shade. When the plants were in sunlight, Ingenhousz saw bubbles forming underneath the leaves. But in the shade, there were no bubbles. Ingenhousz collected the gas from the bubbles and performed more tests. He found that the gas was oxygen and concluded that plants use sunlight to produce oxygen. Through his experiments, Ingenhousz discovered photosynthesis, the process by which green plants change sunlight into chemical energy.

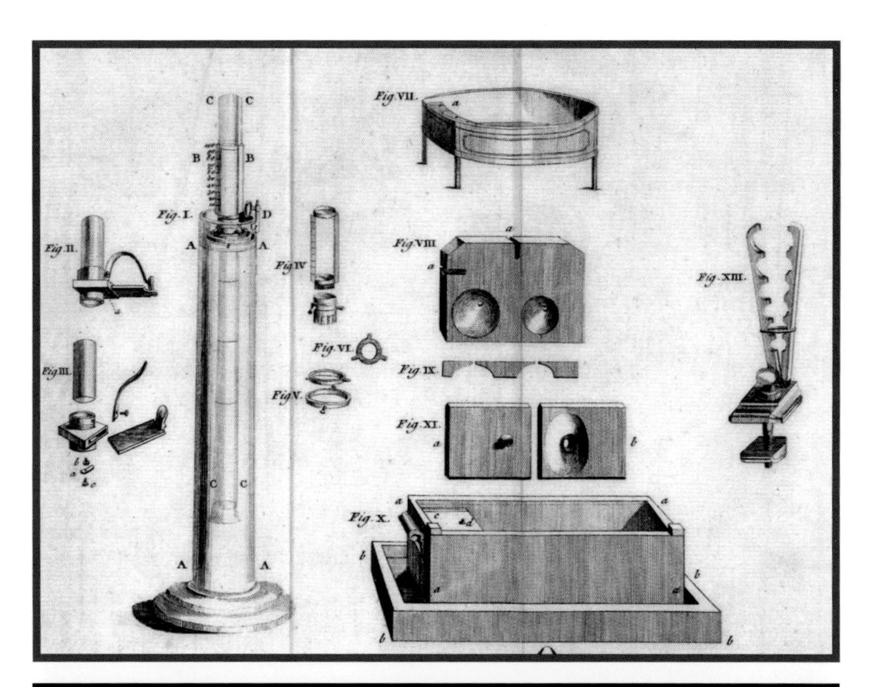

Ingenhousz conducted more than five hundred experiments on plants and published his findings in *Experiments upon Vegetables*. This engraving appears in his book.

CELLULAR ENERGY

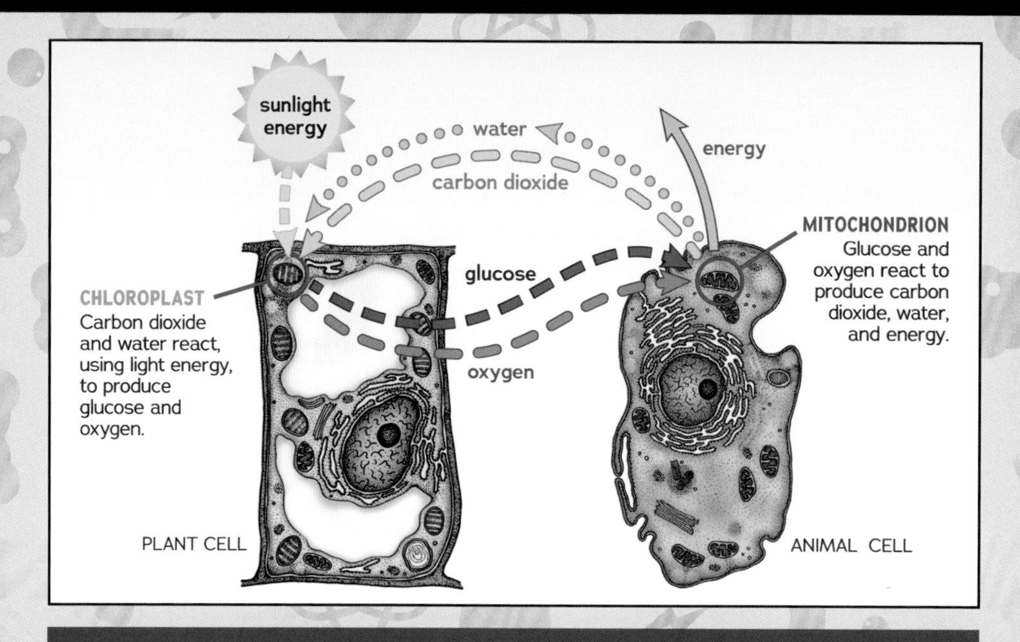

Plant cells receive energy from the sun, which they convert into glucose (sugar) and oxygen. Animal cells receive sugar and oxygen from plants, which they convert into energy.

Inside green plant parts are cells called chloroplasts. They contain a pigment called chlorophyll, which absorbs sunlight. The chloroplasts turn sunlight, water, and **carbon dioxide** into sugar and oxygen. Plants store and use the sugar as their food.

Plants are filled with potential energy stored as chemical energy. Your body can't make chemical energy from sunlight as plants can, but your body's **digestive system** can get the chemical energy locked inside food.

WHAT DOES THE LARGEST ANIMAL EAT?

The blue whale is the largest animal to have ever lived on Earth. But you might be surprised by what it eats. To get the nutrients it needs, blue whales eat tiny shrimplike animals called krill. Krill get their energy from tiny plants that float at the ocean's surface. To get all the energy it needs, a blue whale can eat 4 tons (3.6 metric tons) of krill in one day!

Your digestive system breaks down the food you eat so that cells in your body can use its stored energy.

A chemical reaction happens as food breaks down into smaller and smaller pieces in your digestive system. This reaction changes the food into nutrients your body can absorb and delivers the nutrients to the cells around your body. Your body uses these nutrients as energy to move, grow, and heal.

But how do you get that energy from your cells? In the 1840s, scientists found their first clues about how cells release energy when they discovered rod-shaped parts in the cells of the body. These parts—mitochondria—combine glucose, fat, and protein with oxygen to make chemical energy that the body can use.

CHAPTER 4

WHERE DO WE GET ENERGY FOR MACHINES?

Plants don't just fuel our bodies—they fuel our machines too. Fossil fuels such as coal, natural gas, and petroleum are made from plant remains. When plants die, they fall to the ground and become buried. Over time, heat and pressure from Earth change this plant matter into coal, natural gas, and petroleum.

Burning fossil fuels releases their energy. But people didn't know how to make the most of fossil fuels until Scottish engineer James Watt experimented with machinery in the late eighteenth century. He designed an engine that was three times more efficient than previous machines.

Coal is one of the most inexpensive fossil fuels, but burning and mining coal has a negative impact on the environment.

Soon many industries began producing goods in factories that used fossil fuels.

These days, natural gas heats homes by fueling furnaces. Car engines burn gasoline, which is made from petroleum. Coal is used to make electricity. But Earth has a limited supply of fossil fuels. Once we use all of them, we cannot replace them.

NUCLEAR POWER

A cleaner form of energy, called nuclear power, can be made from tiny particles called atoms. All matter is made from atoms. Energy holds the pieces of atoms together. But scientists can release that energy in a process called fission, which breaks apart atoms of uranium, a heavy metal. Scientists used fission to make electricity for the first time in 1951. In nuclear power plants, the heat released from breaking the atoms makes steam, which turns **turbines** to make electricity.

Americans use hundreds of millions of gallons of gasoline each day. But scientists have been working since the twentieth century to come up with different forms of energy and fuel. Can you think of a way to use less gasoline?

HOW NUCLEAR ENERGY IS PRODUCED

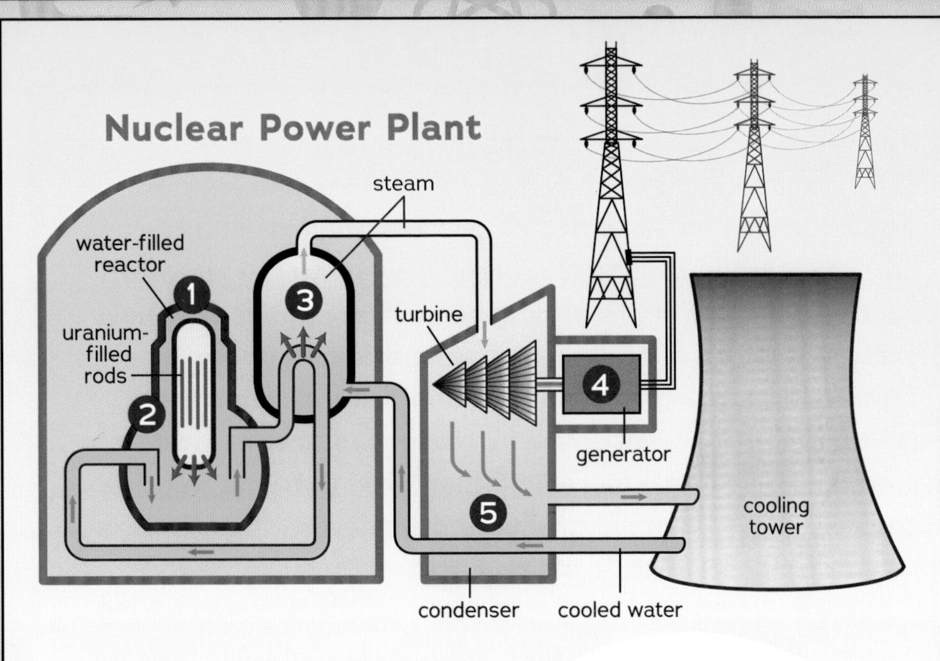

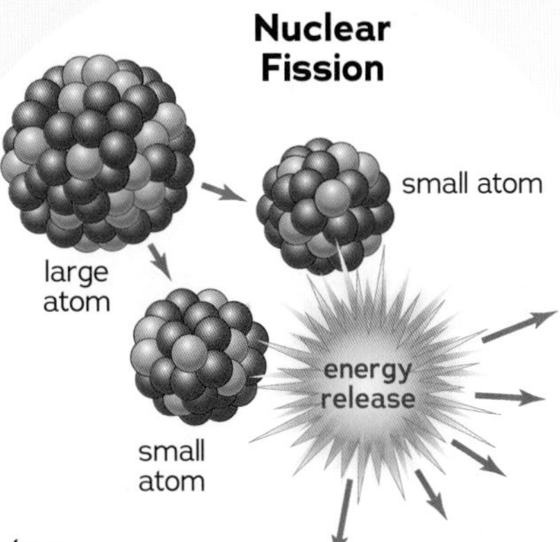

1. Metal rods are filled with uranium pellets. The rods are placed in water in a nuclear reactor.
2. Uranium atoms split and release energy in the rods.
3. The energy heats water and creates steam.
4. The steam moves a turbine. The turbine turns a generator. This creates electricity.
5. The steam cools and turns into water. The water is used again. A cooling tower may be used to release extra heat from the plant.

CHANGING OUR CLIMATE

People have been burning fossil fuels in large amounts since the late eighteenth century, and it has had serious consequences on our environment. Burning fossil fuels releases carbon dioxide and other gases into the air, which become part of Earth's atmosphere. These gases trap heat around Earth's surface, causing temperatures to rise. Earth's climate is changing, and glaciers and polar ice caps are melting. The sea level is rapidly rising, and ocean waters are warmer and contain more acid than ever. Extreme weather causes more rain and extreme temperatures. These human-made changes affect all life on Earth.

CLEAN ENERGY

Renewable energy such as sunlight, wind, and water is the best fuel for the environment. It has an unlimited supply and does not pollute the planet. Wind turbines and **hydropower** from flowing water make electricity. Heat from inside Earth, called **geothermal energy**, can also be used to make electricity.

One early investigator of geothermal energy was Italian Piero Ginori Conti. He saw steam escaping from vents in Earth known as geysers and wondered if he could make electricity from the steam. In 1903 he made the steam move a bladed wheel that powered a machine. Then, in 1904, he was able to produce enough electricity to light five lightbulbs. It was the first time steam from Earth had made electricity.

Wind turbines convert the wind's kinetic energy into mechanical power that can be turned into electricity.

CHAPTER 5

WHAT MAKES A ROLLER COASTER MOVE?

You can see energy transfer in action just by watching a roller coaster. Do you know how roller coasters use energy? First, a chain pulls the cars to the top of a giant hill. The cars gain more potential energy the higher they go. Then this energy switches to kinetic energy as soon as the cars begin to fly down the hill.

Roller coaster cars gain potential energy as the chain pulls them to the top of the first hill.

At the bottom of the first hill, the cars have the most kinetic energy.

It powers the cars to reach the top of the next hill, which is smaller than the first. Then the cars fly up and down the roller coaster track. The roller coaster cars switch from potential energy to kinetic energy and don't even use an engine or fuel to keep moving. As the roller coaster's hills get smaller and smaller, the cars slow down.

Roller coaster cars move around the track simply because of energy.

Did you think energy could be used in so many ways? Thanks to the hard work of scientists over many years, we know that energy is found everywhere on Earth, in many forms—from inside a plant to the motion of a car. It can be stored as potential energy, and it can be used as kinetic energy to get things moving. But there may be more to learn. Von Helmholtz wanted to find out how animals produce heat. He wasn't sure the scientists who had studied it before were right. Mouchot was concerned that the world would run out of coal, and he wanted to find a new way to run machines. What questions do you have? What problems do you want to solve? Like Becquerel or Ingenhousz, you can conduct experiments, changing one thing at a time, until you see something happen. Then you can

What kind of energy does this stretched rubber band have?

Energy keeps us warm, helps us move, and powers our world. We also use energy to have fun!

do more research and testing to understand what you saw and to come up with a conclusion, or an answer to your question. With study, questions, and testing, we can continue learning just how much energy moves through our world.

TRY IT!

You've read about how machines use different kinds of energy to work. Try it for yourself with this experiment. Will rubber bands stretched to different lengths have different amounts of energy?

MATERIALS

- sidewalk chalk
- a long sidewalk or driveway
- rubber bands (same length and kind)
- ruler
- a helper
- tape measure
- paper and pencil

PROCEDURE

1. Draw a starting line on the sidewalk. You will stand there during the experiment.
2. Shoot a rubber band from your starting point. Hook it to the front part of a ruler. Pull it back to the 4-inch (10-centimeter) mark. Let go.
3. Ask your helper to draw a chalk circle around the spot where each rubber band lands.
4. Repeat steps 2 and 3 with two more rubber bands.
5. Measure the distance from your line to each marked circle. Record the distances on a piece of paper.

6. Try stretching a rubber band farther. Shoot it in the same way as before, but pull it back to the 6-inch (15 cm) mark. Repeat with two more rubber bands. Record the distances they flew.
7. Repeat again, but stretch the three rubber bands back to the 8-inch (20 cm) mark. Measure and record their distances.
8. Average the distances for each try: Add the three numbers together. Then divide the total by three to get the average.

REVIEW THE RESULTS

Review your recorded results by looking at the average distances the rubber bands from each set of tries flew. Did pulling the rubber band back farther make the distances longer? Does a rubber band's potential energy increase the more it is stretched? Write down your conclusions.

GLOSSARY

carbon dioxide: a gas released when humans and animals breathe out or when fossil fuels are burned. Plants take in some carbon dioxide.

conserved: saved from being lost or wasted

digestive system: a body system that converts the energy in food to nutrients that the body can use to make energy

electromagnetic radiation: the range of energy in sunlight

geothermal energy: heat from inside Earth

hydropower: electricity made by moving water

reflectors: devices that reflect light, heat, or sound energy

renewable energy: energy made from natural resources with unlimited supplies, such as wind or the sun

turbines: engines that are driven by water, steam, or gas that pass through the blades of a wheel to make it move

FURTHER INFORMATION

BBC Bitesize: Energy Basics
http://www.bbc.co.uk/bitesize/ks3/science/energy_electricity_forces/energy_transfer_storage/revision/2/

Bright, Michael. *From Oil Rig to Gas Pump*. New York: Crabtree, 2016.

Doeden, Matt. *Finding Out about Solar Energy*. Minneapolis: Lerner Publications, 2015.

Environmental Protection Agency: A Student's Guide to Global Climate Change
https://www3.epa.gov/climatechange/kids/index.html

Green, Dan. *Eyewitness Energy*. New York: DK, 2016.

NASA: Climate Kids—Energy
http://climatekids.nasa.gov/menu/energy

Sneideman, Joshua, and Erin Twamley. *Renewable Energy: Discover the Fuel of the Future with 20 Projects*. White River Junction, VT: Nomad, 2016.

US Energy Information Administration: Energy Kids
http://www.eia.gov/kids

INDEX

PHOTO ACKNOWLEDGMENTS

The images in this book are used with the permission of: design elements: iDesign/Shutterstock.com; © iStockphoto.com/kotoffei. Galina Barskaya/Shutterstock.com, p. 4; © Shunsuke Yamamoto Photography/DigitalVision/Thinkstock, p. 5; INTERFOTO/Personalities/Alamy Stock Photo, p. 6; © Laura Westlund/Independent Picture Service, pp. 7, 16, 21; Pictorial Press Ltd/Alamy Stock Photo, p. 8; © biosdi/iStock/Thinkstock, p. 9; © monkeybusinessimages/iStock/Thinkstock, p. 10; © bruev/iStock/Thinkstock, p. 11; © Science Source, p. 12; © GraphicaArtis/Hulton Archive/Getty Images, p. 13; © PavelRodimov/iStock/Thinkstock, p. 14; © Science & Society Picture Library/Getty Images, p. 15; © Patricio Robles Gil/Minden Pictures, p. 17; © Purestock/Thinkstock, p. 18; © Thinkstock Images/Stockbyte/Thinkstock, p. 19; © FeelPic/iStock/Thinkstock, p. 20; © XXLPhoto/iStock/Thinkstock, p. 22; © iStockphoto.com/globestock, p. 23; © iStockphoto.com/Marcio Silva, p. 24; © iStockphoto.com/AntonBalazh, p. 25; © roberthyrons/iStock/Thinkstock, p. 26; © iStockphoto.com/kali9, p. 27.

Front cover: NASA/GSFC/SDO.